System Efficiency

Improvement Opportunities

Introduction

Publisher Integrity Company, LLC

Integrity Institute of Technology

Copyright 2016

Author/Instructor David R. Carpenter, PhD

Copyright Statement/Disclaimer

Table of Contents

Introduction

The goal of an efficient system is to provide reliability and reduce operational cost. In this book you find proven methods of practical applications that produce positive results. Case studies, proven engineering methods, testing and many other ways to determine the best solutions to your existing facility and/or machine are provided in this book. Defining and identifying energy efficiency opportunities is the most cost prohibitive way to improve the profit and loss ratio.

Use this book as a guide to find energy efficiency opportunities. This book is full of illustrations, pictures and diagrams that help define and understand energy efficiency. It has an associated class available online.

Industrial Sector Energy Use Based on NEMA and DOE

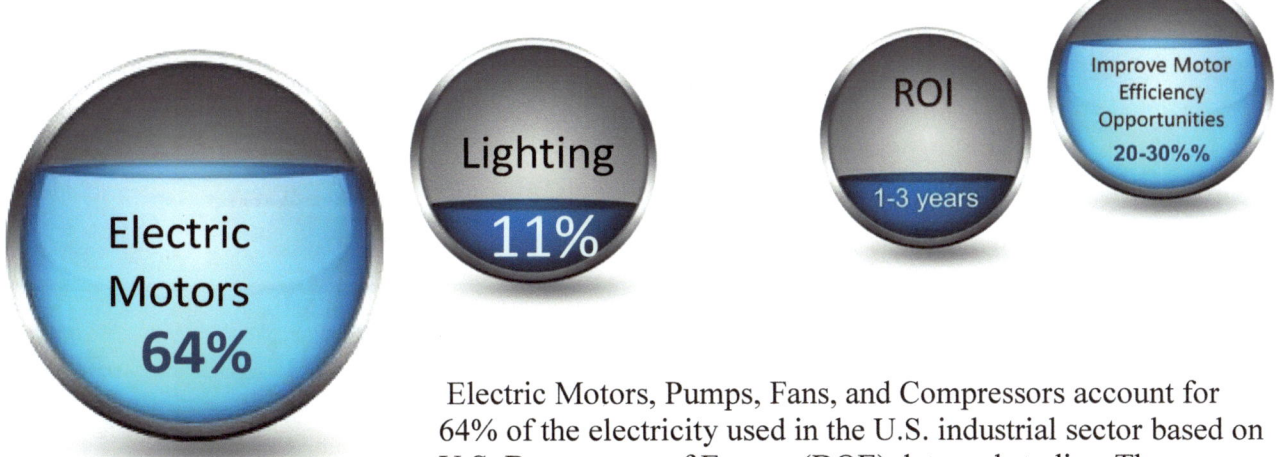

Electric Motors, Pumps, Fans, and Compressors account for 64% of the electricity used in the U.S. industrial sector based on U.S. Department of Energy (DOE) data and studies. The National Electrical Manufacturers Association (NEMA) and DOE estimate that an efficiency motor program would save 5.8 terrawatts of electricity and prevent the release of nearly 80 million metric tons of carbon into the atmosphere over the next ten years. This is equivalent to keeping 16 million cars off the road.

Energy represents more than 97 percent of total motor operating costs over the motor's lifetime. Using best practice, energy efficiency of electrical motors can be improved by 20% to 30% on average. Improvements should have a payback time of 1 to below 3 years.

Based on DOE and NEMA findings around 30 million new electric motors are sold each year for industrial purposes. Some 300 million motors are in use in industry, infrastructure and large buildings. These electric motors are responsible for 40% of global electricity used to drive pumps, fans, compressors and other mechanical traction equipment. Motor technology has evolved over the last few decades.

ROI vs Efficiency and Reliability

*Financial Impact of Motor Consumption and Savings for Selected Industries

Industry Groups	Motor Sys. Costs/ Estab.	Motor Energy Costs / Total Operating Costs	Savings per Estab. Per Year	Savings as % of Operating Margin
Paper Mills	4.6 million	6.5%	$659,000	5%
Petroleum Refining	5.6 million	1.4%	$946,000	1%
Industrial Inorganic Chemicals, nec.	1.6 million	10.4%	$283,000	6%
Paperboard Mills	3.0 million	6.4%	$492,000	5%
Blast Furnaces and Steel Mills	6.0 million	2.1%	$358,000	2%
Industrial Organic Chemicals, nec.	1.3 million	1.0%	$91,000	1%
Industrial Gases	1.1 million	21.7%	$116,000	13%
Plastics Materials and Resins	1.5 million	1.5%	$121,000	1%
Cement, Hydraulic	2.2 million	9.6%	$219,000	4%
Pulp Mills	1.7 million	6.7%	$483,000	5%

*Manufacturers Energy Consumption Survey, Bureau of Economic Analysis, Census of Manufacturers and United States Industrial Electric Motor Systems Market Opportunities Assessment

This class/book will include DOE performed case studies with proven methods to reduce electric motor energy consumption with simple and small cost to industry. Also given will be resources and guidance for how, when and why increase electric motor efficiency.
Motors that are not properly managed can and do result in billions of dollars in wasted energy and operating costs to industry. Electric motor-driven systems used in U.S. industrial process

industries consumed 679 billion kWh of electrical energy. Motors used in industrial space heating, cooling, and ventilation systems use an additional 68 billion kWh. A detailed analysis of the U.S. motor systems inventory indicates that this energy use could be reduced by 11% to 18% if plant managers implement all cost-effective applications of mature and proven energy efficiency technologies.

Case Studies

Sources: Motor Challenge DOE, Manufacturers Energy Consumption Survey, Bureau of Economic Analysis, Census of Manufacturers, and United States Industrial Electric Motor Systems Market.

Opportunities Assessment

Energy represents more than 97 percent of total motor operating costs over the motor's lifetime. Using best practice, energy efficiency of electrical motors can be improved by 20% to 30% on average. Improvements should have a payback time of 1 to below 3 years.

Table E-3: Summary of Motor Challenge Showcase Demonstration Projects

Company	Type of Plant	Energy Savings kWh/Year	Savings as % of Initial Sys. Energy	Annual Cost Savings	Payback on Investment (Years)
General Dynamics	Metal fabrication	451,778	38%	$68,000	1.5
3M Company	Laboratory facility	10,821,000	6%	$823,000	1.9
Peabody Coal	Coal processing	103,826	20%	$6,230	2.5
Stroh Brewery	Beer brewing	473,000	52%	$19,000	0.1
City of Milford	Municipal sewage	36,096	17%	$2,960	5.4
Louisiana-Pacific	Strand board	2,431,800	50%	$85,100	1.0
City of Trumbull	Sewage pumping	31,875	44%	$2,614	4.6
Nisshinbo California	Textiles	1,600,000	59%	$100,954	1.3
Greenville Tube	Stainless steel tubing	148,847	34%	$77,266	0.5
Alumax	Primary aluminum	3,350,000	12%	$103,736	0.0
OXY-USA	Oil field pumping	54,312	12%	$5,362	0.5
City of Long Beach	Waste incineration	3,661,200	34%	$329,508	0.8
Bethlehem Steel	Basic oxygen furnace steel mill	15,500,000	50%	$542,600	2.1
Total/Average		38,663,734	33%	$2,166,330	1.5

Case Studies

☐ Chevron, completed motor system efficiency improvement project at its Richmond, California plant, resulting in $700,000 savings annually

☐ Weyerhaeuser- saves $2.5 Million 1st year, 2.5 million second year with Motor Efficiency plan according to John R. Homquist, PE, Life Fellow IEEE

Kodak Rochester, New York

Kodak's TMP – Total Motor Program Saves -$664,000 annually, Reduces annual energy consumption more than 5.8million kWh, Reduces maintenance costs, and ROI 2.3-year simple payback. In 1995, staff at Kodak's plant in Rochester, New York, launched Kodak's Total Motor Program (TMP) to consolidate the plant's inventory of motors and associated spare parts. Later, the focus of the TMP was expanded to improving energy efficiency by retrofitting the plant's existing motors with smaller, more efficient motors.

Since 2002, Kodak has retrofitted approximately 600 motors with National Electrical Manufacturers Association (NEMA) Premium™ efficiency models.

Cost savings from these retrofits are 5,802,000 kWh and $664,000, respectively. Since total project costs were approximately $1.5 million, the simple payback is slightly more than 2 years.

Lessons Learned: Aging from and improperly configured industrial motor systems can waste energy and increase maintenance and operating costs

Minnesota Mining and Manufacturing (3M)

Minnesota Mining and Manufacturing (3M) conducted an in-house motor system performance optimization project. Using a systematic facility-by-facility approach, the company formed a team that evaluated approximately 1,000 electric motor systems in 29 buildings at the 3M Center to identify feasible projects. Four key energy-saving upgrades in Building 123 reduced electricity use by 41 percent and resulted in cost savings of $77,554 per year. The systematic approach developed, and experience gained in the Building 123 project, was applied to other 3M facilities and demonstrates how a large industrial company can optimize performance of their electric motor systems at a campus-type facility.

ANNUAL ENERGY AND COST SAVINGS

	Validated Projects		Entire Campus (est.)	
Electricity Savings	939,400 kWh	$31,583	10,821,000 kWh	$363,800
Utilities (reduce steam and chilled water use)		$45,971		$441,200
Maintenance Savings				$18,000
Total		$77,554		$823,000

Air to Motor Case Studies

Department of Energy's Industrial Assessment Center located at the University of Alabama at Tuscaloosa, Neptune Industries, Tallassee, Alabama

Table 1. Opportunities at Neptune				
Recommended Action	**Annual Resource Savings**	**Annual Cost Savings ($)**	**Implementation Costs ($)**	**Payback (years)**
Replace Metal Halide Fixtures	170,057 kWh/yr	$10,438	$23,295	2.2
Utilize Photocell Sensors	47,869 kWh/yr	2,938	1,003	0.3
Reduce Compressed Air System Pressure	498,643 kWh/yr	30,417	1,000	0.0
Reduce Leaks in Compressed Air System	874,808 kWh/yr	53,363	3,900	0.1
Eliminate Bag-house fan	581,398 kWh	35,465	27,200	0.8
Install Automatic Louvers and Relocate Thermostats	3,734 MMBtu/yr	28,528	3,584	0.1
Totals	**1,543,508 kWh/yr**	**$161,149**	**$59,982**	**0.4**

By simply reducing leaks annual operational cost were reduced by $30, 417 + $53,363 = $83,780 total. Reducing leaks prevents the motor/compressor from running too often but also prevents the motor from wasting energy during normal operation. This also extends the life of the compressor/motor.

Feedback Learning

1. Electric Motors account for 64% of the electricity used in the U.S. industrial sector based on U.S. Department of Energy (DOE) data and studies. The National Electrical Manufacturers Association (NEMA) and DOE estimate that an efficiency motor program would save 5.8 terrawatts of electricity and prevent the release of nearly 80 million metric tons of carbon into the atmosphere over the next ten years. This is equivalent to keeping 16 million cars off the road. **T** or F

2. Energy represents more than 97 percent of total motor operating costs over the motor's lifetime. Using best practice, energy efficiency of electrical motors can be improved by 20% to 30% on average. Improvements should have a payback time of 1 to below 3 years. **T** or F
   a. Develop Efficiency Program
   b. Utilize Proven Methods That Develop Efficiency Program
   c. Understand the Relationship Between Reliability and System Efficiency
   d. <u>All the above</u>

3. Maintenances and engineering play an important role to reliability. **T** or F
4. Maintenances and engineering play an important role to efficiency. **T** or F
5. Efficiency cannot exists without reliability. **T** or F
6. Bethlehem Steel saves $542,600 a year after making basic maintenance and engineering corrections which increased efficiency. **T** or F

```
┌─────────────────────────────────────────┐
│                                         │
│                                         │
│                                         │
│         Electric Motor Performance       │
│                                         │
│                                         │
│                                         │
└─────────────────────────────────────────┘
```

Key Solution Components to Develop Energy Efficiency:

1. Develop an Energy Efficiency Program
2. Utilize Proven Methods That Develop Efficiency Program
3. Understand the Relationship Between Reliability, Safety, Facility and System Efficiency
4. Train the Engineers, Technicians and Management who are the key people that can make or break an efficiency program.

Why Motors and Machines Are Excellent Opportunities to Optimize Energy Efficiency

When motors operate frequently at low loads and over a wide range of conditions, there are often many excellent opportunities to optimize the entire system, save energy, and improve reliability by making various improvements.

Improvement opportunities can include replacing the motor with one of a more appropriate size or type, or installing a speed-adjusting device (or both) based on Operating Hours and Full load of Motor and torque matching. Improper quality of power feeding the motor or machine such as voltage unbalanced, too low or too high can waste power and shorten motor life because the motor is in operational stress.

In considering whether to downsize a motor, it is important to check the load duty cycle to avoid overloading the motor during peak-load conditions. This is especially applicable in seasonal industries that experience peak loads only a few times each year.

Motors inherently operate at a loss (see below) because of the push pull action, see example below. For more information of how motors work see our book on Electric Motors at www.integrityco.com

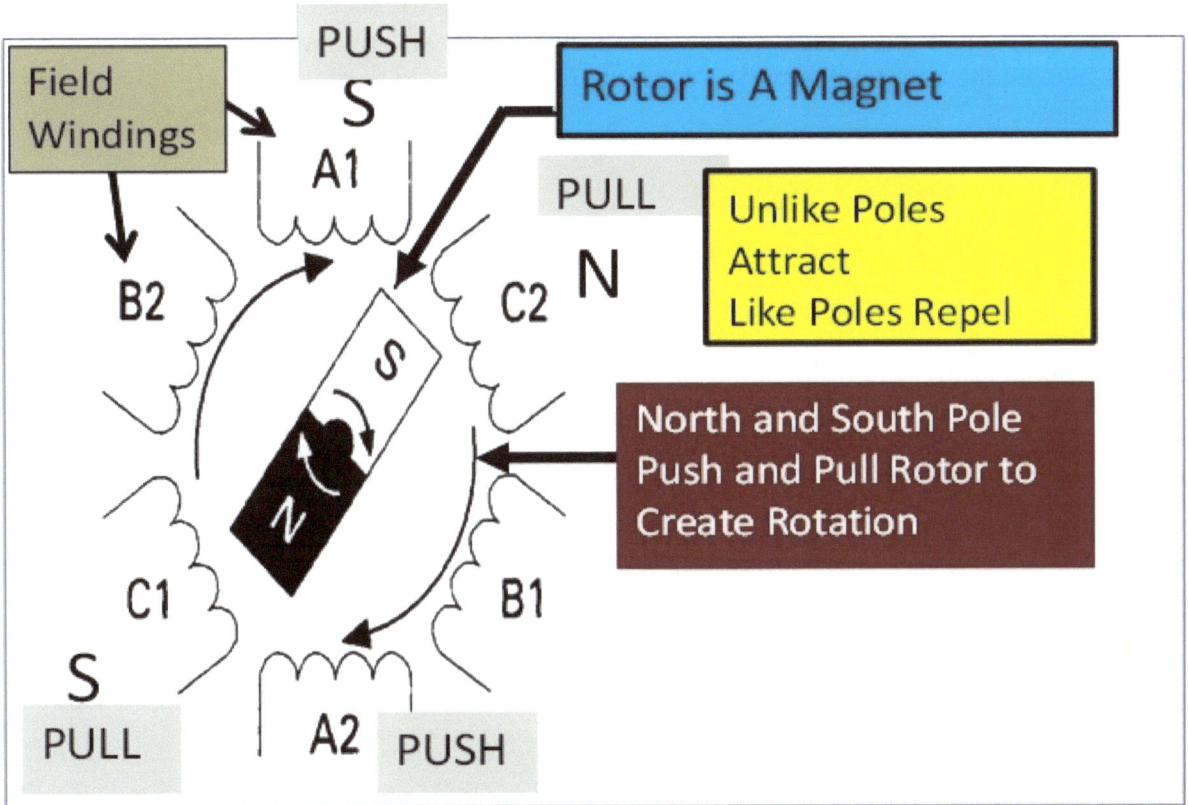

Common Motor Efficiency Consideration "Nuts and Bolts"

- ☐ Inherently Operate at loss
- ☐ Motor is Dependent on System Efficiency
- ☐ Inefficient when operating 40% of the rated load
- ☐ Motors operate most efficient at about 70% to 80% of load
- ☐ Oversized Motors
- ☐ Poor Power Factor
- ☐ Poor Power Quality
- ☐ Undersized Motors
- ☐ Voltage

Field Windings   **Rotor**

The Problem with Oversized Motors

Engineers frequently specify motors that are larger than needed to meet system requirements in order to ensure that the existing motor/drive assembly can support anticipated increases in capacity. However, the consequences of oversizing motors include the following:
• Lower efficiency
• Higher motor/controller costs
• Higher installation costs
• Lower power factor
• Increased operating costs.

Motor operation at low load factor can result when the driven load is smaller than anticipated. In many applications, original equipment manufacturers will overstate the horsepower needs of their equipment to avoid liability in specifying a motor that cannot meet service requirements. This practice and the tendency of engineers who assemble the system to be conservative can lead to the selection of a motor that operates far below its rated capacity.

How to Determine Energy Waste during Normal Operation

KVA represents power that is given by the power provider. KW represents the real or true power that is actually used in the system. KVAR represents wasted energy. This energy is wasted by normal operation.

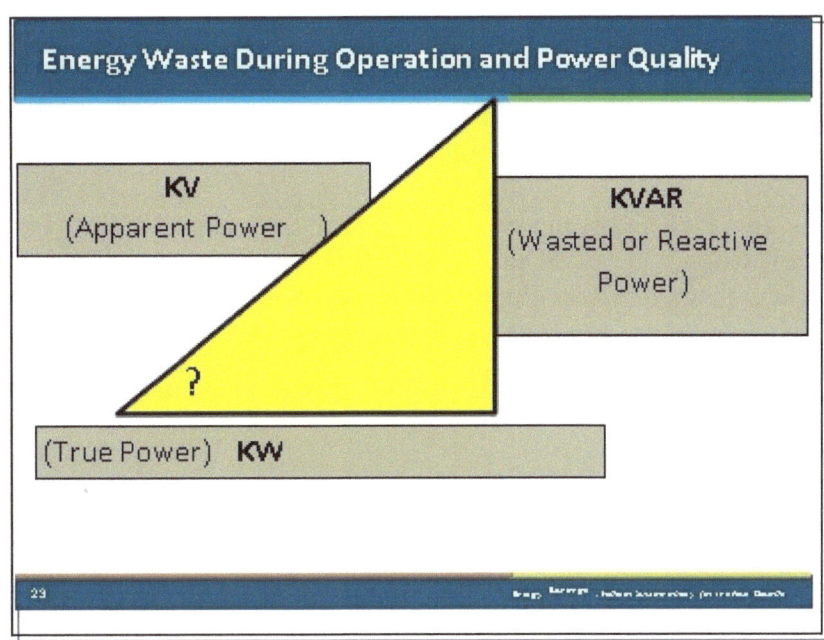

The Problem with Undersized Motors

Motors should be sized to operate from 75% to 100% of rated load. The principal consequence of operating a motor above its rated load is a higher winding temperature, which shortens the operating life of the motor and waste energy.

If the motor has a service factor of 1.0, the motor lifetime may last only a few months if it is operated above rated load or if it is operated at rated load when there is a power quality problem.

- Motors should be sized to operate from 75% to 100% of rated load
- Increases Winding Temperature
- Service Factor
- Undersize motors waste energy during operation

The Problem With Voltage Unbalanced Motors

As a rule of thumb, every 10°C rise in winding temperature reduces insulation life by half. Although motor efficiency drops off slightly at higher-than-rated loads, the increase in energy cost is usually not as severe as the cost associated with shorter intervals between repairs or replacements.

Load balancing is critical to stopping wasted energy. The illustration to the right shows equally balanced loads and should be our target.

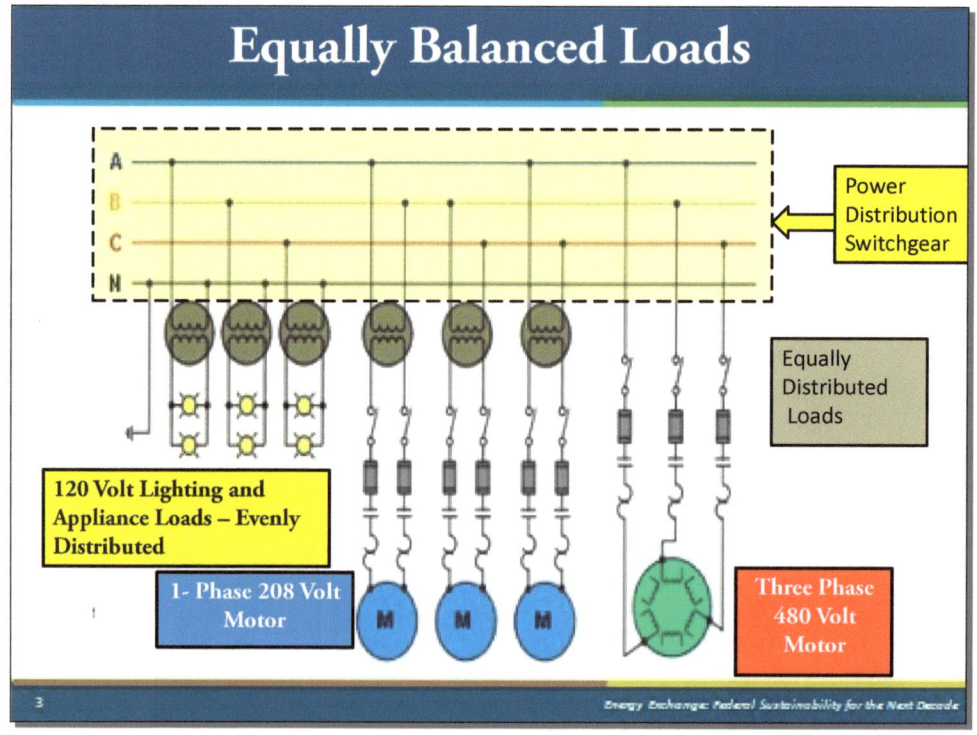

Effects of unbalanced loads are shown. This installation is a major waster of energy because it causes the motors to operate under stress. The continual stress not only waste energy but also shortens the life of the motor and will affect other associated equipment.

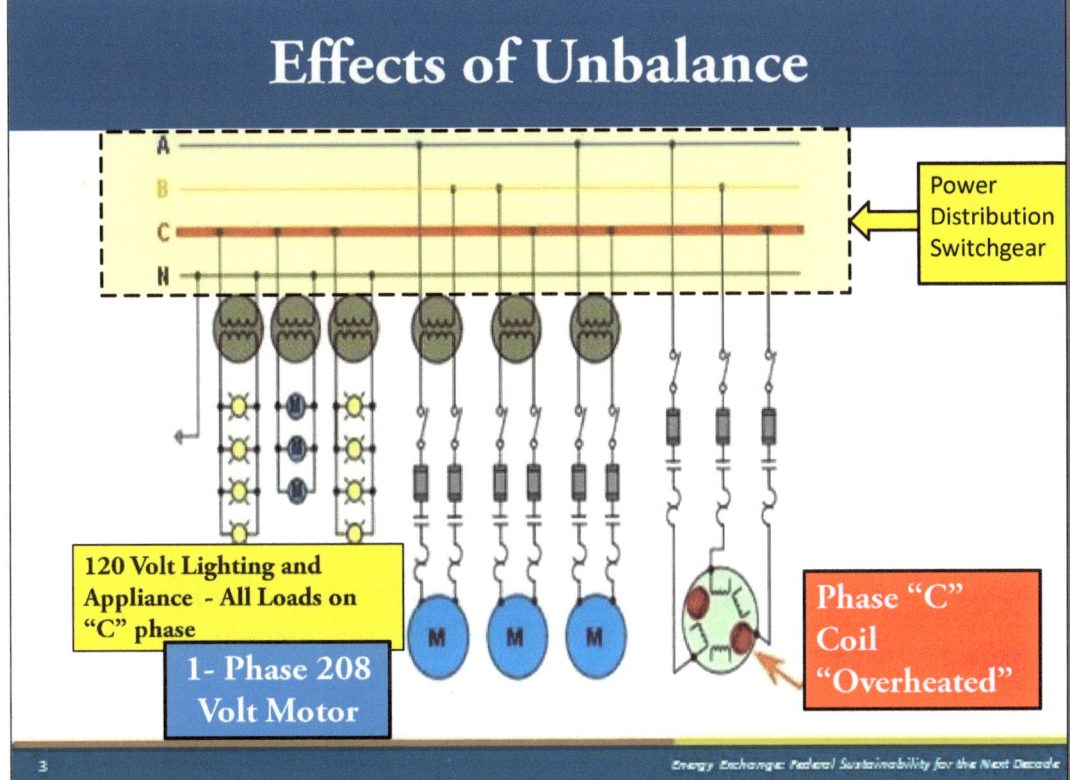

Hot spots such as shown in the illustration to the right are caused normally by unbalanced voltage, harmonic currents or unbalanced loads. When this motor is running it is wasting a tremendous amount of energy. Simple test can be used to determine this problem. An appropriate Power Analyzer would be useful to determine this problem.

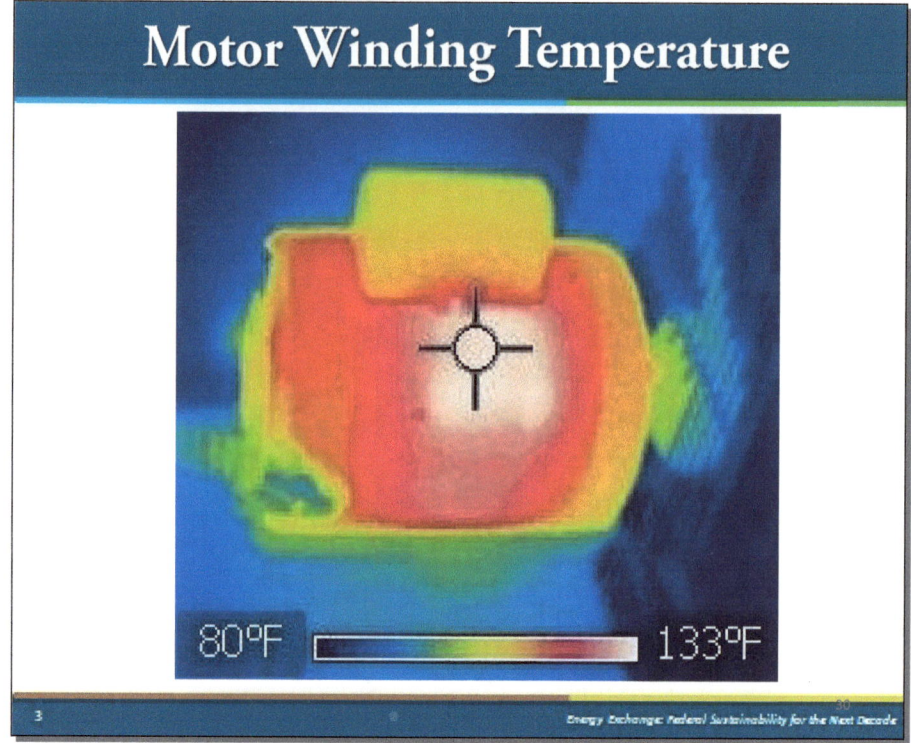

Motor Winding Temperature

80°F   133°F

Wasted Use Analysis

To the right is an example of a motor which has been wasting energy for one year. The cause is unbalanced voltage. The wasted energy is 30,240 KWh per year.

Voltage Drops (1) Electric Motor for (1) Year

Circuit	Measured Voltage Drop, Volts	Excess Voltage Drop, Volts	Current, Amps	Excess Power, kW	Excess Energy Use, kWh/year
L₁	8.1	5.6	199.7	1.12	9,796
L₂	5.9	3.4	205.7	0.7	6,126
L₃	10.6	8.1	201.8	1.63	14,318
			Totals:	3.45	30,240

In the illustration to the right we see major heat gathering at "C" phase of this disconnect. A closer look at the disconnect reveals that the "C" phase has been replaced because it is different from the other. This could be caused by a ground fault condition or overload condition. In this case, the "C" phase has unbalanced voltage caused by improper wiring methods.

Fused Disconnect

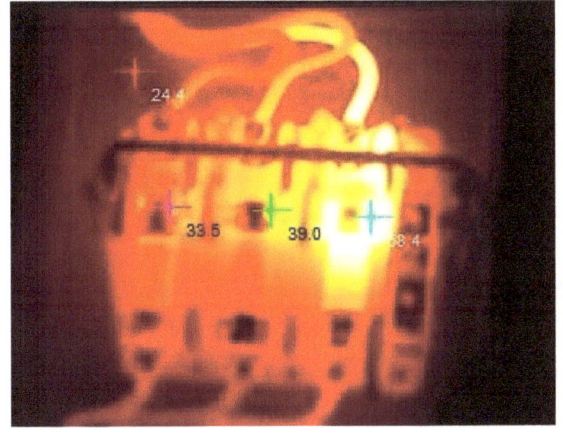

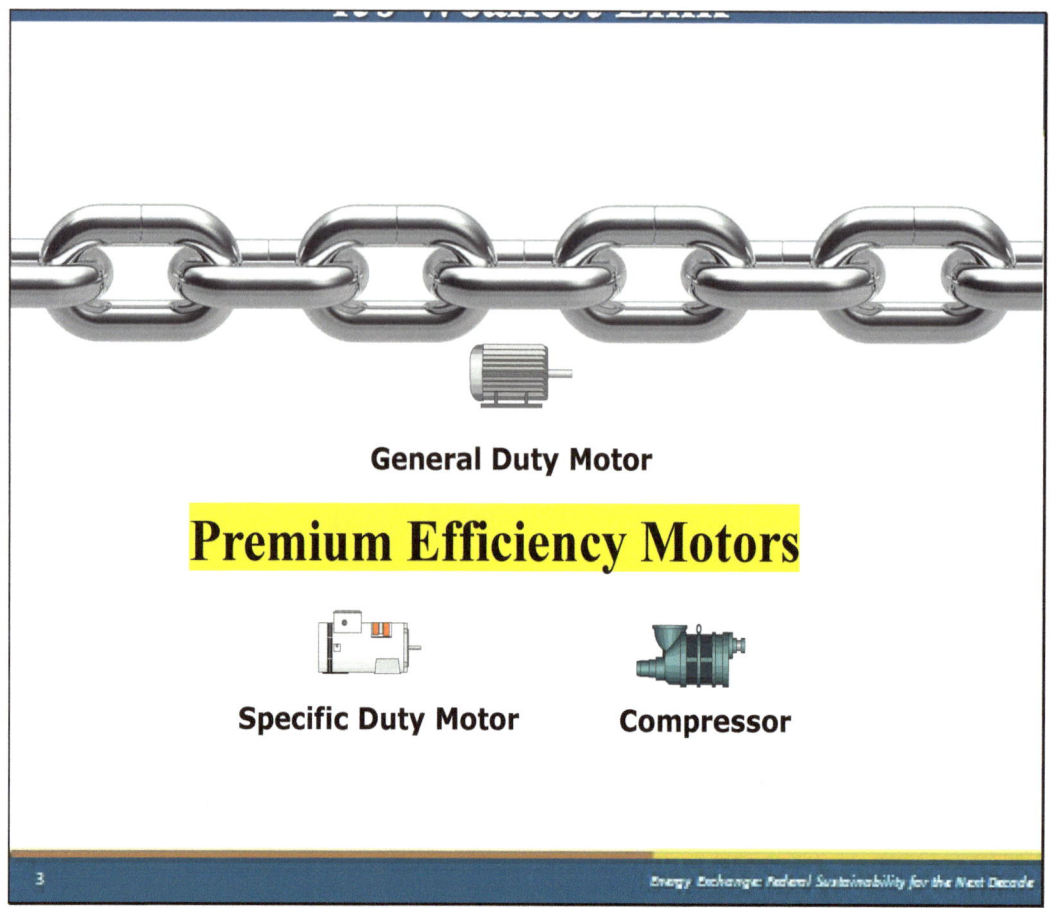

Best Production is only as Good as its Weakest Link

System Efficiency Determines All Aspects of Efficiency

✓Power Distribution –Switchgear, Panels, Transformers, Wiring Methods, Overcurrent Protection, Grounding , Bonding and Shielding.
✓Installation and Design
✓Motor, Pump, Fan and Compressor Efficiency
✓Motor Controls (Drives & PLC's) – maintenance, design and installation
✓Mechanical Matching
✓Input Output Balance

An example of a poorly performing system involves an oversized combustion air fan serving a wood waste burner. The combustion air flow is controlled by an outlet louver damper. Efficiencies of individual components are critical to overall performance

Example: Poor Performing Electrical/Mechanical System Has An Oversized Combustion Air Fan Serving A Wood Waste Burner

- ❑ Fan = 55% (the fan is not operating at its best efficiency point)
- ❑ Drive = 96%
- ❑ Motor = 90% (the fan is equipped with an old standard efficiency motor)
- ❑ Control = 28% (damper losses)
- ❑ Install = 89%% (reduction in fan efficiency due to poor installation practices)
- ❑ Power Distribution = 75% (assumed)
- ❑ **System Efficiency = 0.55 × 0.96 × 0.90 × 0.28 × 0.89 × 0.75 = 0.0888 or .08.8%**
- ❑ Note: improving the **motor efficiency from 90% to 95%** would improve system efficiency from 8.8% to 9.04%

3

Energy Exchange: Federal Sustainability for the Next Decade

Motor to Pump Efficiency

η stands for the efficiency of a power supply, defined as the output power divided by the input power.

Every thermodynamic system exists in a particular state. A thermodynamic cycle occurs when a system is taken through a series of different states, and finally returned to its initial state. In the process of going through this cycle, the system may perform work on its surroundings, thereby acting as a heat engine.

A heat engine acts by transferring energy from a warm region to a cool region of space and, in the process, converting some of that energy to mechanical work. The cycle may also be reversed. The system may be worked upon by an external force, and in the process, it can transfer thermal energy from a cooler system to a warmer one, thereby acting as a refrigerator or heat pump rather than a heat engine.

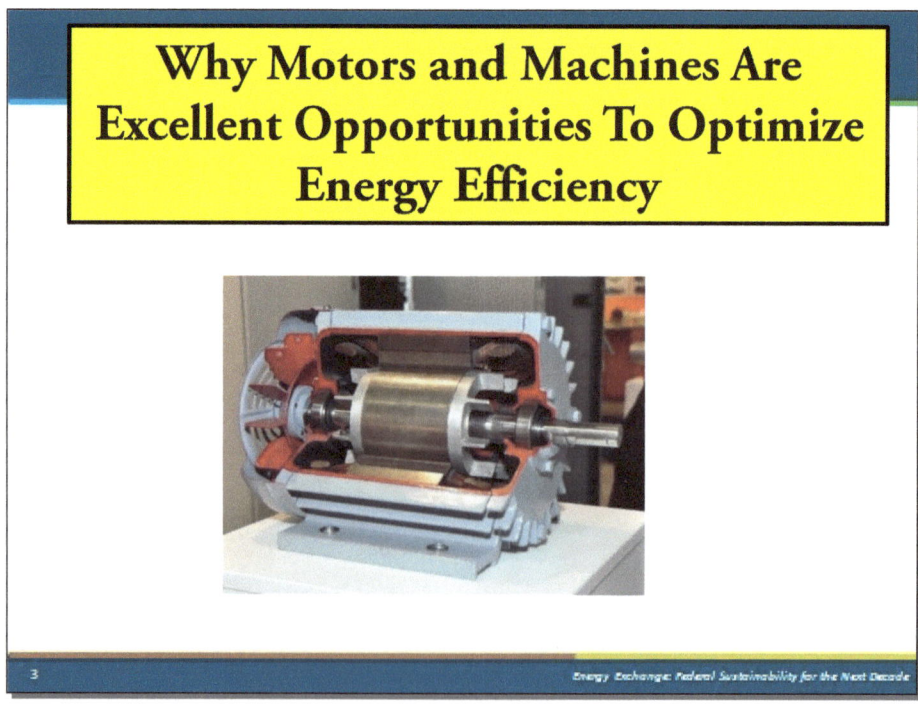

Example:

η Fan = 55% (the fan is not operating at its best efficiency point)

η Drive = 96%

η Motor = 90% (the fan is equipped with an old standard efficiency motor)

η Control = 28% (damper losses)

η Install = 89%% (reduction in fan efficiency due to poor installation practices)

ηDistribution = 98% (assumed)

η System = 0.55 × 0.96 × 0.90 × 0.28 × 0.89 × 0.98 = 0.116 or 11.6%

For this example, improving the motor efficiency to 95% would only improve the system efficiency from 11.6% to 12.2%.

When motors operate frequently at low loads and over a wide range of conditions, there are often many excellent opportunities to optimize the entire system, save energy, and improve reliability by making various improvements.

Improvement Opportunities can include replacing the motor with one of a more appropriate size or type, or installing a speed-adjusting device (or both) based on Operating Hours and Full load of Motor

In considering whether to downsize a motor, it is important to check the load duty cycle to avoid overloading the motor during peak-load conditions. This is especially applicable in seasonal industries that experience peak loads only a few times each year.

Common Causes of Motor Failure

✓Poor Power Quality

✓Unbalanced Voltage

✓High Voltage

✓Low Voltage

✓Improper Wiring Method

✓Misapplication of Overcurrent Protection

✓Misuse – Wrong type motor for load

✓Misapplication to load

✓Unsuitability for Operating Environment

✓Misalignment

✓Vibration

✓Poor Maintenance Practice

How to Accomplish Motor Energy Efficiency

How To Accomplish Motor Energy Efficiency

- Understand How the Motor Works
- Perform Analysis of Existing Motor Applications
- Develop Policy for Motor Replacement
  - Replace
  - Repair
- Develop Plan for Replacement of Motors
- Maintain Power Quality to the Motor
- Train Engineers and Technician
- Analyzing Motor Efficiency Opportunities

Motor and Drive System Basics

Overview

Electric motors, taken together, make up the single largest end use of electricity in the United States. In industrial applications, electric motors account for roughly 60% of electricity consumption; in the process industries, electric motors account for more than 70% of electricity use.

Electric motors provide efficient, reliable, long-lasting service, and most require comparatively little maintenance based on maintenance, design and application.

Despite these advantages, however, they can be inefficient and costly to operate if they are not properly selected and maintained. Industrial plants can avoid unnecessary increases in energy consumption, maintenance, and costs, by selecting motors that are well suited to their applications and making sure that they are well maintained.

For example, see illustration to the right, suppose that a motor-driven pump supplies water to several heat exchangers and has a flow requirement that the system piping and heat exchangers were designed to handle. The pump was specified according to the requirements of this flow condition. However, actual operating conditions can vary according to the season, the time of day, and the production rate. To handle the need for variable flow rates, the system is equipped with valves and bypass lines. This equipment can be useful if it is properly applied, but wasteful if it is not.

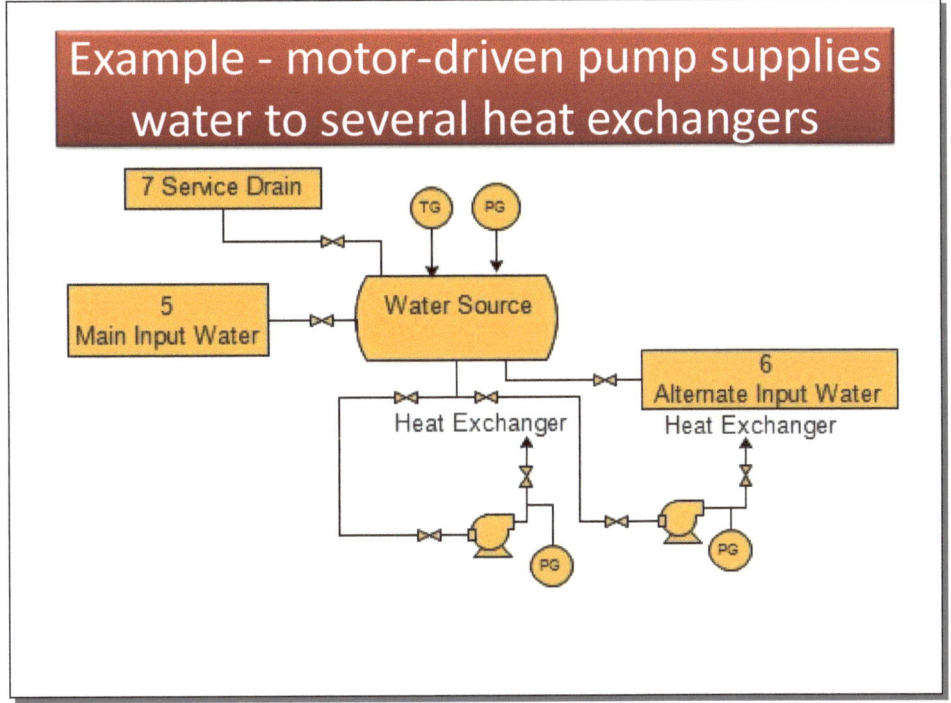

Example - motor-driven pump supplies water to several heat exchangers

Similarly, many fan systems have variable air delivery requirements. A common practice is to size the fan so that it meets the highest expected load and use dampers to restrict airflow during periods of low demand. However, one of the least efficient methods of controlling flow is to use dampers. Consequently, although the system provides adequate airflow, the lack of a drive to control the motor's speed (and thus airflow) can cause system operating costs to be significantly higher than necessary.

In addition to increasing energy costs, an inefficient motor and drive system often increases maintenance costs. When systems do not operate efficiently, the stress on the system caused by energy losses must be dissipated by piping, structures, dampers, and valves. Additional system stresses can accelerate wear and create loads for which the system was not originally designed. For example, in a pumping system, excess flow energy must be dissipated across throttle valves or through bypass valves, or it must be absorbed by the piping and support structure. As a result, all of this equipment can degrade more easily. Throttle and bypass valves can require seat repair, and piping and support structures can develop cracks and leak as a result of fatigue loads. Repairing or replacing this equipment can be costly. A drive could resolve this issue with good return of investment.

In addition, inefficient system operation in an industrial plant can create poor working conditions, such as high levels of noise and excessive heat. High noise levels can be the result of flow noise, structural vibrations, or simply operating the equipment. Excessive noise can fatigue workers more quickly and thus reduce productivity.

In addition, inefficient systems often add heat to the workplace. This added heat usually must be removed by the facility's heating, ventilating, and air-conditioning (HVAC) system, further increasing total operating costs.

Indications of Poor System Design

Taking a component-based approach to industrial system design and operation tends to increase facility costs and maintenance requirements and reduce reliability. However, the problems associated with a poorly designed system—high energy costs, the need for frequent maintenance, and poor system performance—can be corrected, as indicated below.

1. High Energy Costs

High energy costs can be the result of inefficient system design as well as inefficient motor operation. Not selecting or designing a proper motor and drive system for the application can also lead to power quality problems, such as voltage sags, harmonics, and low power factor.

2. Frequent Maintenance

Equipment that is not properly matched to the requirements of the application tends to need more maintenance. The primary causes of increased maintenance requirements are the added stresses on the system and the increased heat that accompanies inefficient operation. Ironically, system designers often specify oversized motor and drive and end-use equipment in order to improve reliability. An oversized motor might be more reliable, but it might also make other parts of the system less reliable. A more effective way of ensuring high reliability is to design a system and specify system components so that the system's operating efficiency is good/balanced over the full range of operating conditions.

Indications of Poor System Design

1. High Energy Costs
2. Frequent Maintenance
3. Poorly System Performance
4. Voltage Sags
5. Harmonics
6. Bad Power Factor
7. Equipment Is Improperly Matched To Load Over/Under Sized

3. Poor System Performance

Operating a motor and drive system that was not properly selected for its application can result in poor overall system performance. Poor system performance is a major cause of increases in maintenance and decreases in reliability. Common indications include abrupt or frequent system starts and stops, high noise levels, and hot work environments. In many material handling systems, the work-in-process moves roughly from one work station to the next. The banging that often accompanies sudden accelerations and decelerations is symptomatic of stress on the motor and drive system. The consequences of this stress can be more frequent maintenance and poor operating efficiency.

High noise levels are common in inefficient fluid systems. Since energy losses in fluid flow often dissipate as noise, systems with large flow losses tend to be loud. In addition, inefficient equipment operation often greatly increases the temperature of the workspace, especially if the added heat load was not included in the design specifications for the HVAC system.

4. Voltage Sags

This is discussed at length later. Sag are when the voltage falls and rises usually during a high demand time. Sags cause the system to waste energy and create stress on the system. Probable causes are normally improper wiring methods, connections or other associated loads.

5. Harmonics

Harmonics are discussed later in depth. Harmonics are produce by loads that switch on different points of the wavelength and create even and odd wave forms synchronized with the fundamental wave. They produce unwanted currents that can cause overheating and waste energy.

6. Bad Power Factor

Bad power factor is a sign of extreme waste of energy. This is discussed in depth later. Most utilities will penalize, in some form, the customer for bad power factor.

7. Improperly Matched Equipment

When motors, pumps, fans and compressors are improperly matched they produce stress and wasted energy. This is discussed later in depth.

1. System Efficiency is only as good as it's weakest link. **T** or F
2. US Department of Energy has estimated that half of all compressed air is wasted. **T** or F
3. Do Motor Failures Effect Energy Efficiency? Yes or No
4. Chronic Motor Failure Occurs because of which of the Following:
    - A- Control Poor Power Quality
    - B- Develop Plan for Replacement of Motors
    - C- Maintain Power Quality to the Motor
    - D- Train Engineers and Technician
    - E- Analyzing Motor Efficiency Opportunities
    - F- A, B & C
    - G- C,D & E
    - **H- All the Above**
5. Electric Motors Account For 60% Of Electricity Consumption in Process Industries. T or **F**

Types of Electric Motors

To ensure that motors are applied properly, it is essential to understand the various types of motors and their operating characteristics. Electric motors fall into two classes, based on the power supply: alternating current (ac) or direct current (dc). The most common types of industrial motors are shown.

Alternating current (ac) motors can be single-phase or polyphase. In terms of quantity, single-phase motors are the most common type, mainly because many small motors are used for residential and commercial applications in which single-phase power is readily available. However, several operating constraints on these motors limit their widespread use in industrial applications.

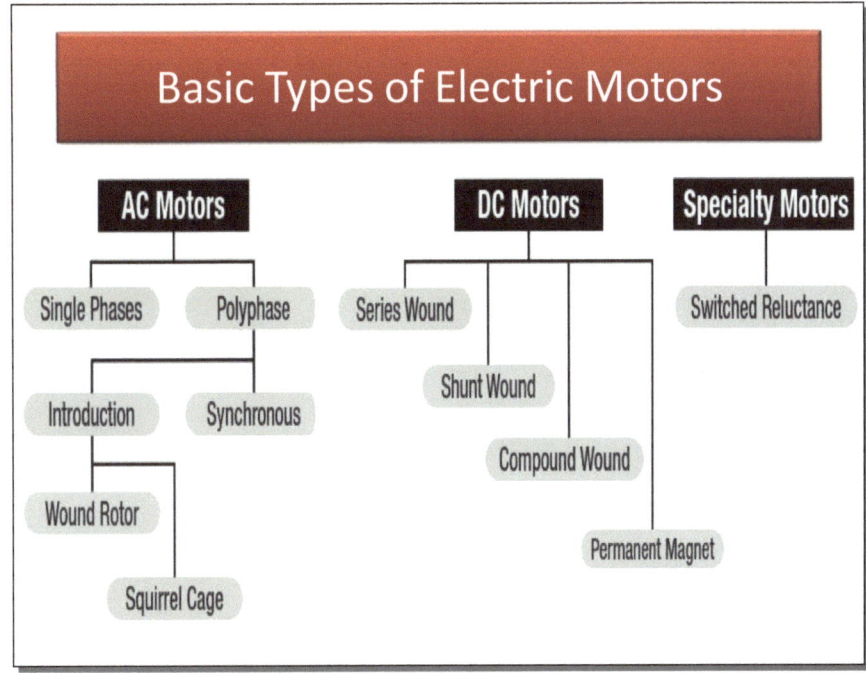

Integral single-phase induction motors tend to pull large starting currents relative to the motor's size. In general, they operate less efficiently than three-phase motors of comparable size, and are not available in larger sizes.

In contrast, polyphase motors are used widely in industrial applications. They consume more than half of all the electricity used in industry. Polyphase motors can be found in almost every industrial process, and they often operate continuously to support production processes. These motors can achieve high efficiencies with favorable torque and current characteristics. The effectiveness and low cost of three-phase motors are major reasons why three-phase power is used so widely in industry. In terms of energy consumption and efficiency improvement opportunities, three-phase motor systems predominate.

Assessing Economic Feasibility

Because of better design and low loss/high quality materials, premium efficiency motors typically cost 15% to 30% more than their energy efficient counterparts. In many situations (e.g., new motor purchases, repairs, or motor replacement) you quickly recover this price premium through energy cost savings. To determine economic feasibility, examine the total annual energy savings in relation to the full or incremental cost of purchasing and installing the premium efficiency motor. See the motor energy savings calculation form.

Analyze Motor Efficiency Opportunities

- Cost Analysis
  - Pre Eff motors typically cost 15% to 30% more than standard
  - Rewind motors – Standard Cost
- New motor purchases, repairs, or motor replacement
- See Motor Energy Savings Calculation Form
- See Motor Repair vs Motor Replacement Booklet

Most industrial plant managers base their energy efficient equipment purchase decisions on a simple payback analysis, and they require that investments be recovered through energy savings within 1 to 3 years. The simple payback is defined as the period of time required for the savings from an investment to equal the initial or incremental cost of the investment. For initial motor purchases or replacement of burned-out and non-rewindable motors, the simple payback period for the investment in a new premium efficiency motor is the incremental cost for the premium efficiency motor (less any available utility rebate) divided by the total annual cost savings. No installation costs are assessed as either the premium efficiency or energy efficient motor must be installed.

For the motor repair/replace decision, the simple pay-back is the total cost of the new premium efficient motor minus the repair cost and any utility incentive (if available), divided by the total annual electrical energy and demand reduction cost savings. Motor removal and installation costs are not considered as the failed motor must be removed and a replacement spare or premium efficient motor installed.

For replacement of in-service or operating motors, the simple payback is the ratio of the full cost of purchasing and installing a new premium efficiency motor relative to the value of the total annual electrical savings. Base or "bare" motor installation costs must include an overhead and profit multiplier when outside contractors are used.

A labor cost adjustment should be applied to motors with restricted access or special handling requirements. The simple payback given replacement of an operable motor is given in equation next page.

For the motor repair/replace decision, the simple payback is the total cost of the new premium efficient motor minus the repair cost and any utility incentive (if available), divided by the total annual electrical energy and demand reduction cost savings. Motor removal and installation costs are not considered as the failed motor must be removed and a replacement spare or premium efficient motor installed.

For replacement of in-service or operating motors, the simple payback is the ratio of the full cost of purchasing and installing a new premium efficiency motor relative to the value of the total annual electrical savings. Base or "bare" motor installation costs must include an overhead and profit multiplier when outside contractors are used.

A labor cost adjustment should be applied to motors with restricted access or special handling requirements. The simple payback given replacement of an operable motor is given in Equation above.

Simple Payback for the Replacement of an Operable Motor

$$SPB = ((COST + COST\,INSTALL) - REBATE) \div SAVINGS$$

Where:

SPB = Simple payback in years
Cost = New motor cost
Cost INSTALL = Installation cost
Rebate = Utility rebate for premium efficiency motor
Savings = Total annual cost savings

The following analysis for purchasing a *new 250-hp* TEFC motor operating at 75% of full rated load illustrates how to use Equations. The analysis determines the cost-effectiveness of purchasing a new premium efficiency motor having a 3/4-load efficiency of 96.2% ($\eta = 96.2$) instead of an energy efficient motor (η)EE PREM = 95.5%). The motor is expected to be in operation for 8,000 hours per year.

Electrical energy is purchased at a rate of $0.08/kWh with a demand charge of $8.00/kW-month.

Kilowatts saved:

kW REDUCTION
= 250 × 0.75 × 0.7457 ×
(100/95.5 - 100/96.2) = 1.06 kW

This is the amount of power conserved by the premium efficiency motor during each hour of use. Multiply this by the number of operating hours at the indicated load to obtain annual energy savings.

Energy saved:

kWh SAVINGS
= 1.06 × 8,000 = 8,480 kWh/year Assuming utility energy and demand charges of $0.08/kWh and $8.00/kW-mo. (from Equation 6-3): **Cost Savings = (1.06 kW × 12 mo × $8.00/ kW-mo) + (8,480 kWh/year × $0.08/kWh) = $780/year**

In this example, installing a premium efficient motor reduces the utility bill by $780 per year. The simple payback for the incremental cost associated with a premium efficiency motor purchase is the ratio of the price premium or incremental cost to total annual cost savings. Generally, premium efficiency motors might cost up to 15% to 30% more than a motor of an energy efficient design.

Assuming a price premium of $2,500 and no utility incentive, the simple payback on investment is as follows:

SPB = ($2,500 – 0) / $780 = 3.2 years The additional investment required to purchase a premium efficiency motor is recovered within 3.2 years.

Premium efficiency motors often pay for themselves rapidly through reduced energy consumption and costs. After this initial payback period, annual savings will continue to be reflected in lower operating costs, and they will add to a company's total profits.

Although the energy and cost savings associated with purchasing a premium efficiency motor can be impressive in many applications, selecting the premium efficiency unit is not always cost-effective. Motors that are lightly loaded or infrequently used—such as motors driving control valves—may not consume enough electricity to allow the premium efficiency model to produce significant energy and cost savings.

Remember, for a motor operating under a constant load, the electricity savings associated with an efficiency improvement are directly proportional to annual hours of operation. Special and definite purpose motors may carry a substantial price premium or may not be available in premium efficiency models.

Power Quality Considerations

For a cyclically <u>alternating electric current</u>, RMS is equal to the value of the <u>direct current</u> that would produce the same power dissipation in a <u>resistive load</u>.

The crest factor of an AC current waveform is the ratio of waveform's peak value to its rms value: crest factor = |peak current| / rms current

The crest factor for a sinusoidal current waveform is 1.414 since the peak value of a true sinusoid is 1.414 times the rms value. Current waveforms for purely resistive loads are sinusoidal, so the crest factor will be 1.414. <u>Some loads, such as switching power supplies or lamp ballasts, have current waveforms that are not sinusoidal.</u> They draw a high current for a short period of time, and their crest factors, therefore, can be quite a bit higher than 1.414.

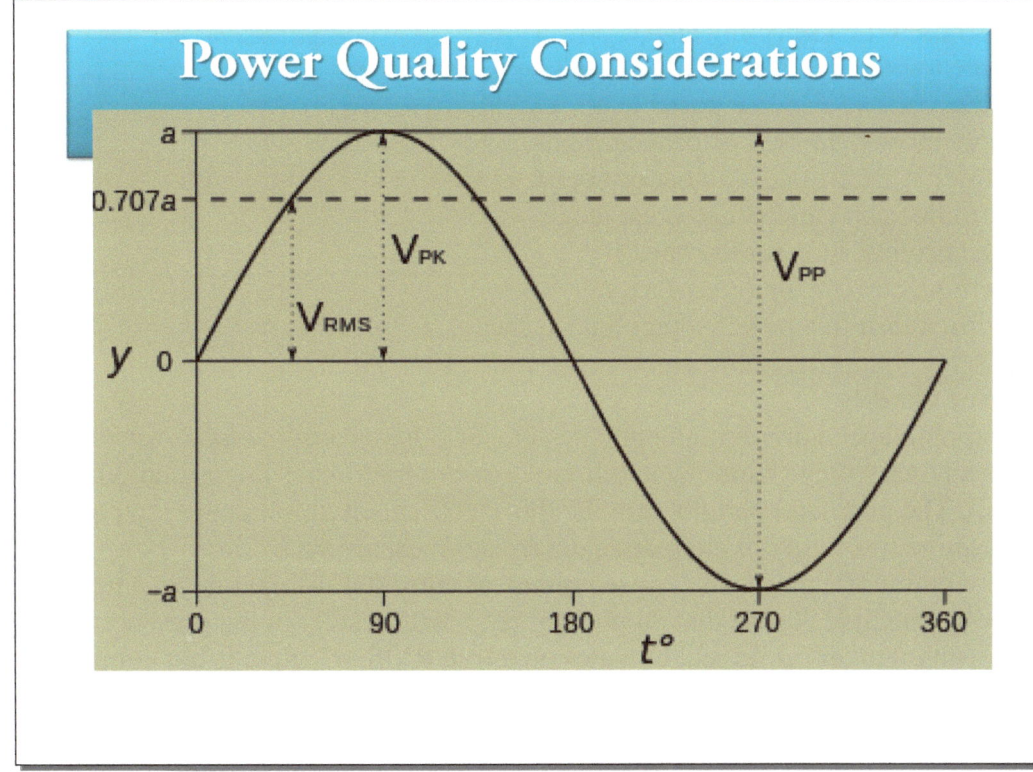

Shown are current waveforms for two different loads, one sinusoidal (the blue trace) and one non-sinusoidal (the red trace). Both have an rms current of 5A, but as you can see the crest factor is quite different.

These two waveforms both have an rms current of 5 A, but their crest factors are very different.

As you would expect, the sinusoidal current waveform has a crest factor of 1.414:
crest factor = |peak current| / rms current = 7.07 A / 5 A = 1.414

The non-sinusoidal current waveform, on the other hand, has a peak value of 21.21 A. The crest factor for this waveform is then:
crest factor = |peak current| / rms current = 21.21 A / 5 A = 4.24

Why is crest factor important?

Both loads draw the same amount of true power (assuming that the input voltage is the same for both). This means that a power source selected to feed the loads at 120VAC would need to provide the 600VA that both loads require.

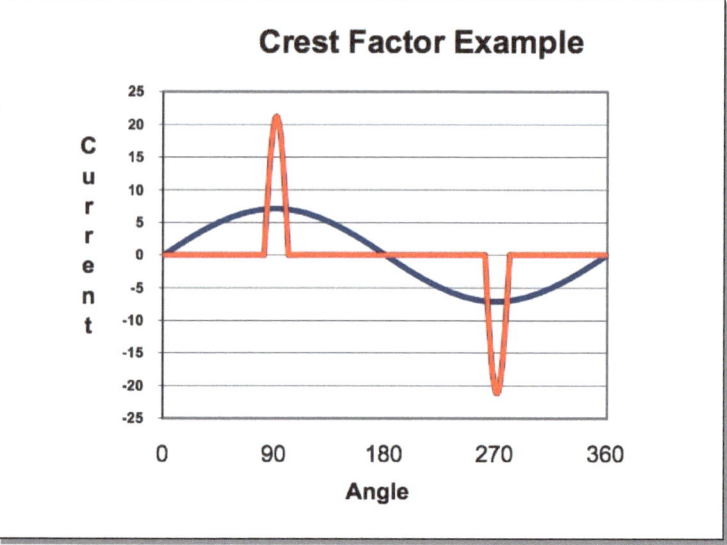

A power source with a power output rating of 600VA may not, however, be able to provide the required peak currents that the non-sinusoidal load demands. When selecting an AC power source to power this load, you would need to choose a source possible of supply more than 21 A of peak current. In order to determine whether or not an AC source can handle high crest factor peak currents, look for the "peak repetitive current" or "crest factor" specifications in a power source's spec sheet.

Power Quality

If variable-speed drives, induction heaters, or other electronic loads are on the system, expect the presence of voltage harmonics. Extreme current harmonics are present in circuits feeding these loads. The purchaser should describe this electrical environment to the equipment supplier and determine the ability of alternative devices to measure such "dirty" power accurately.

At a minimum, devices that sense voltage or current must operate on a true root mean square (RMS) principle. Those that do not will read inaccurately in the presence of harmonics. Knowing that an instrument operates on a true RMS principle is not completely sufficient, however as all instruments are limited by the magnitude and frequency of the harmonics they can handle while still reading accurately. One index of this capability is the crest factor, which is the ratio of peak value to RMS value of a wave form. A perfect sine wave has a crest value of 1.414. True RMS instruments should have a crest factor of 3.0 or better.

Unfortunately, crest value alone is not a complete descriptor of harmonic content. Harmonics caused by most electronic loads cause the current crest factor to be higher and the voltage crest factor to be lower than sinusoidal. An instrument needs to be able to measure accurately across the frequency range of the harmonics. The frequency of harmonics is expressed either in hertz (Hz) or in the order of harmonic. To convert order to frequency, simply multiply by 60. Most high-quality electrical testing instruments specify the frequency range over which their accuracy is maintained.

Three Phase Power Systems

These systems are discussed in depth in our book "Fundamentals of Electrical Design - Transformers

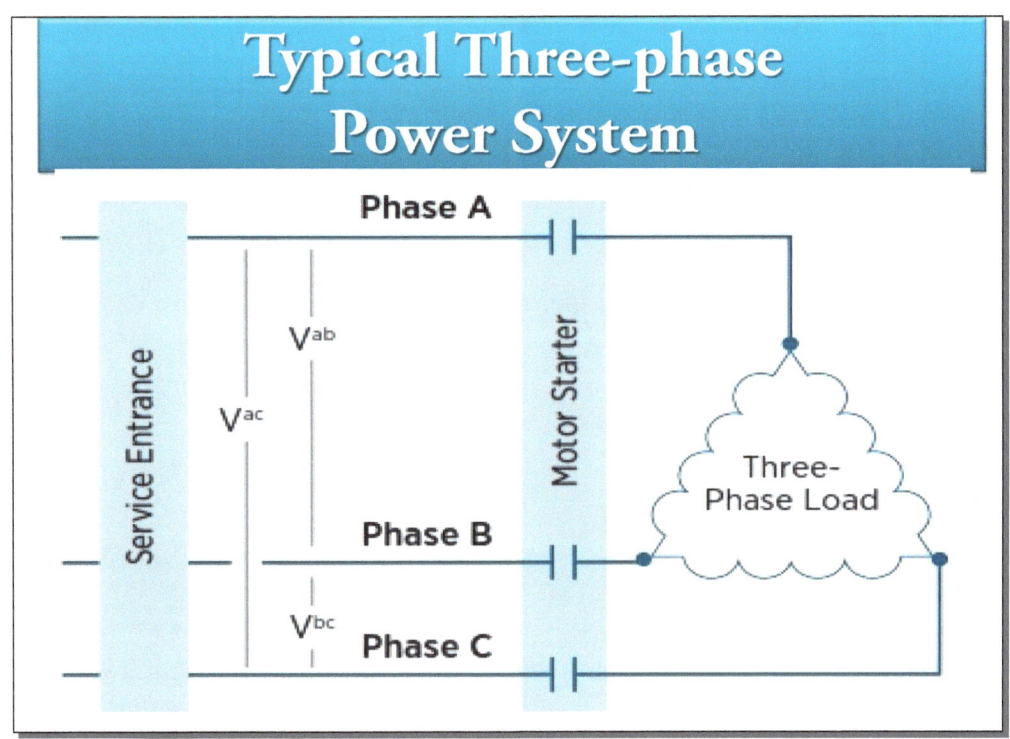

Power Analyzers

Power analyzers are valuable tools to determine the heartbeat of your system. If properly used they can tell you the when energy is wasted as well as find problems when troubleshooting. Keep in mind that there different types of equipment and you must match the proper test equipment to your system to obtain accurate results. This equipment may give indications of a grounding problem are not design as grounding type of equipment.

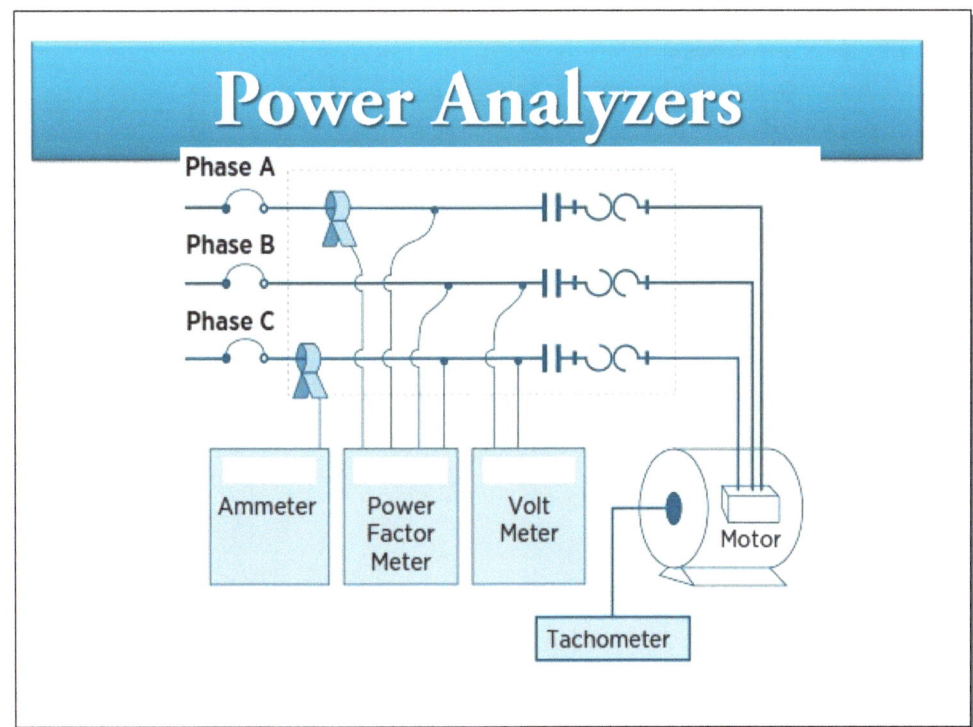

Terminology

Apparent Power (kVA) - is determined by multiplying I (current) x E (voltage) for single phase applications times the voltage. In a three-phase circuit, multiply the average phase-to-phase voltage times the average line current times the square root of 3 divided by 1,000. The units are kilovolt-amperes (kVA).

APPARENT POWER $= V \times I \times \sqrt{3} \div 1,000$

Power Factor Control
• Apparent Power (kVA)
• Reactive Power (kVAR)
• True Power (KW)

Reactive Power (kVAR) This term describes the magnetizing requirements of an electric circuit containing inductive loads. The value of magnetizing power is determined by multiplying the apparent power (kVA) by the sine of the phase angle, θ, between the voltage and the current. Units are kilovolt-amperes reactive (kVAR).

REACTIVE POWER = APPARENT POWER $\times$ sine θ

True or Real or Working Power (kW) This term is used when referring to plant loads. Real power is related to apparent power by the cosine of the phase angle, between voltage and current. Units are kilowatts (kW).

TRUE POWER (KW) = APPARENT POWER $\times$ cosine θ or APPARENT POWER $\times$ PF

Power Factor Terminology
• What is Power Factor?
– Apparent Power
– True Power
– Reactive Power or Wasted Power

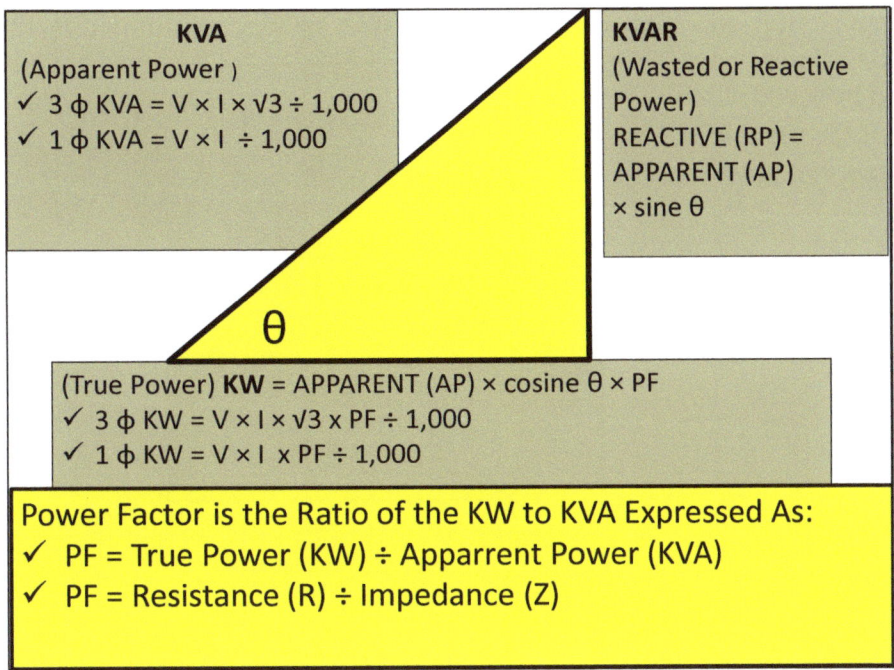

KVA
(Apparent Power)
✓ 3 φ KVA = V × I × √3 ÷ 1,000
✓ 1 φ KVA = V × I ÷ 1,000

KVAR
(Wasted or Reactive Power)
REACTIVE (RP) = APPARENT (AP) × sine θ

θ

(True Power) **KW** = APPARENT (AP) × cosine θ × PF
✓ 3 φ KW = V × I × √3 x PF ÷ 1,000
✓ 1 φ KW = V × I x PF ÷ 1,000

Power Factor is the Ratio of the KW to KVA Expressed As:
✓ PF = True Power (KW) ÷ Apparrent Power (KVA)
✓ PF = Resistance (R) ÷ Impedance (Z)

Overview
Power factor is a measure of how effectively electrical power is being used. A high power factor (approaching unity) indicates efficient use of the electrical distribution system; a low power factor indicates poor use of the system.

Many loads in industrial electrical distribution systems are inductive. Examples include motors, transformers, fluorescent lighting ballasts, induction furnaces and generally non-linear loads. The line current drawn by an inductive load has two components:

Reactance - magnetizing current and power-producing current. The magnetizing current is the current required to sustain the electromagnetic flux or field strength in the machine. This component of current creates reactive power that is measured in kilovolt-amperes reactive, or kVAR. Reactive power does not do useful "work," but it circulates between the generator and the load. It places a heavy drain on the power source, as well as on the distribution system of the power source.
The real (working) power-producing current is the current that reacts with the magnetic flux to produce the mechanical output of the motor.

Real (true) power is measured in kilowatts and can be read on a wattmeter. Real (working) power and reactive power together make up apparent power. Apparent power is measured in kilovolt-amperes or kVA. Real or True power is the actual power consumed.

Another way to visualize power factor and demonstrate the relationship between kW, kVAR, and kVA is the right "power" triangle illustrated, see next page. The hypotenuse of the triangle represents the apparent power (kVA), which is the system voltage, multiplied by the amperage times the √3 (for a three-phase system) divided by 1,000.

The right side of the triangle represents the reactive power (kVAR).

The base of the triangle represents the real or working power (in kW). The angle between the kW and the kVA legs of the triangle is the phase angle θ.

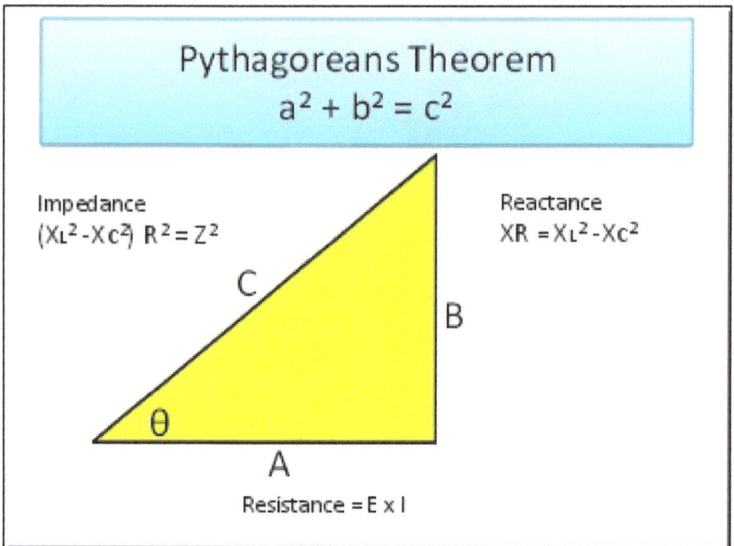

Power factor is the ratio of real power to apparent power. To determine power factor (PF), divide real power (kW) by apparent power (kVA). In a sinusoidal system, the power factor is also referred to as the cosine θ.

Example

Problem: If a sawmill has a Metered demand of 1500 Kw (True Power) and the Measured (Apparent Power) Is 2000 KVA

Solution: Divide 1500KW by 2000KVA to obtain a Power Factor of 0.75. The Phase Angle Is Arc Cosine .75 or 41.4 degrees

Capacitive loads are leading (current leads voltage), and inductive loads are lagging (current lags voltage).

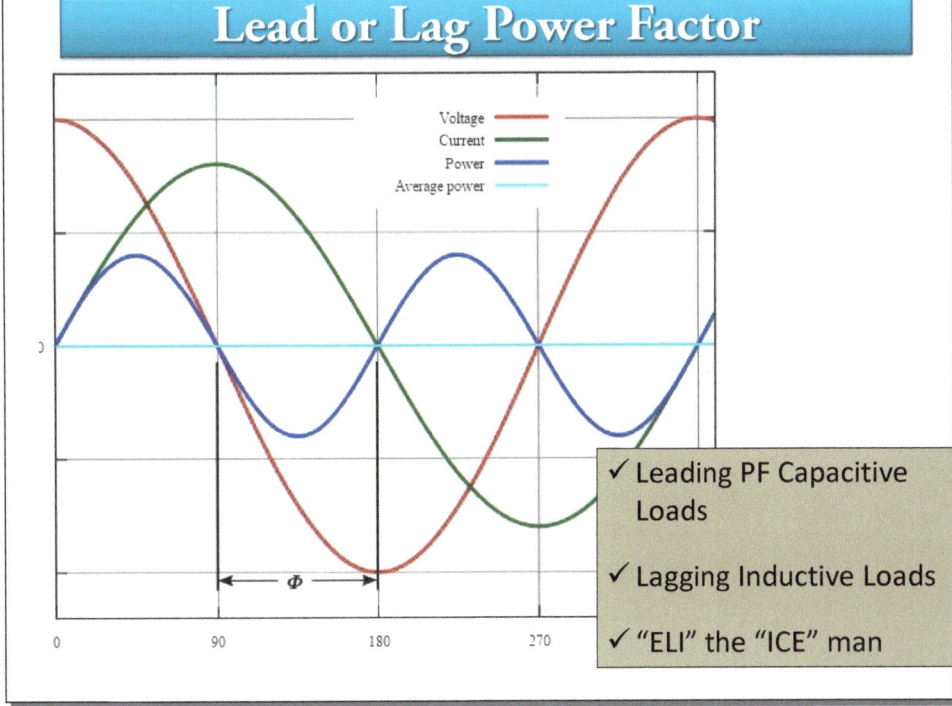

If the resulting current phase angle is more negative in relation to the driving (source) voltage phase angle, then the power factor is said to be "lagging".

If the resulting current phase angle is more positive in relation to the driving (source) voltage phase angle, then the power factor is said to be "leading".

So if the driving voltage phase angle is \theta deg and the resulting current phase angle is \phi deg.

If \theta > \phi power factor is lagging.

If \theta < \phi power factor is lagging.

Then if \theta = \phi power factor is unity and neither leading nor lagging.

The driving (source) voltage phase is often assumed to be zero (for convenience) and in that situation it is immediately obvious that a lagging power factor condition is indicated by a negative sign for the current phase angle. Similarly a positive sign for the current phase angle indicates a leading power factor

The blue line shows all the power is stored temporarily in the load during the first quarter cycle and returned to the grid during the second quarter cycle, so no real power is consumed

Power factor is also referred to as leading or lagging. In the case of the magnetizing current, the power factor is lagging in that the current follows the voltage waveform.
The amount of lag is the electrical phase angle between the voltage and the current.
Displacement power factor is equal to the cosine of the phase angle between the voltage and current waveforms.

Power Factor Penalties

When a utility serves an industrial plant that has poor power factor, the utility must supply higher current levels to serve a given load. In a situation in which real power demand (kW) at two plants is the same, but one plant has an 85% power factor while the other has a 70% power factor, the utility must provide 21% more current to the second plant to meet the identical real power requirement.

Conductors and transformers serving the second plant would need 21% more carrying capacity than that provided to the first plant. Additionally, resistance losses (IR losses) in the distribution conductors in the second plant are increased by 46%.

A low power factor may also lower your plant's utilization voltage, increase electrical distribution system line losses, and reduce the in-plant distribution system's capacity to deliver electrical energy.

A utility is paid primarily on the basis of energy consumption (kWh) and peak monthly demand (kW). Without A power factor billing element, the utility would receive no more income from the second plant than it does from The first. As a compensation for supplying the extra current, utilities often establish a "power factor penalty" .

NOTE: A minimum power factor value is established, often set at 95%. When the customer's power factor drops below the minimum value, the utility collects "low power factor" revenue.

Power Factor Improvement

Induction motors are generally the principal cause of low power factor because many motors are in use that are not fully loaded. When motors operate near their rated load, the power factor is high. For lightly loaded motors, however, the power factor drops significantly.

This effect is partially offset as the total current is less at reduced load.

Lower power factor does not necessarily increase the peak kVA demand because of the accompanying reduction in load.

Power factor can also be improved through replacement of standard efficiency with premium efficiency motors which are appropriately matched to their driven loads. Power factors vary considerably based on motor design and load conditions. While some premium efficiency motor models offer power factor improvements of 2 to 5 percent, others have lower power factors than typical equivalent standard motors. Even motors with high power factor are affected significantly by variations in load. A motor must be operated near its rated loading in order to realize the benefits of a high power factor design.

Power factor can also be improved and the cost of external correction reduced by minimizing operation of idling or lightly loaded motors and by avoiding operation of equipment above its rated voltage. While motor full-and part-load power factor characteristics are important, they are not as significant as nominal efficiency.

When selecting a motor, conventional wisdom is to purchase based on efficiency and correct for power factor.

Some strategies for improving your power factor follow:

• Use a motor with the highest speed that an application can accommodate. Two-pole (nominal 3,600 RPM) motors have the highest power factors; power factor decreases as the number of poles increases.

• Choose motor sizes that are as close as possible to the horsepower demands of the load. A lightly loaded motor requires little real power. A heavily loaded motor requires more real power. Since the reactive power is almost constant, the ratio of real power to apparent power varies with induction motor load, and power factor ranges from about 10% at no load.

Feedback Learning

1. Power factor has nothing to do with Energy Efficiency. T or <u>F</u>
2. Power Factor is a measure of how effectively electrical power is being used. <u>T</u> or F
3. Low power factor (approaching unity) indicates efficient use of the electrical distribution system T or <u>F</u>
4. High power factor indicates poor use of the system. T or <u>F</u>
5. Motors, transformers, fluorescent lighting ballasts, and induction furnaces are considered as inductive loads. <u>T</u> or F
6. Reactance is real power T or <u>F</u>
7. True Power is wasted power T or <u>F</u>
8. Apparent Power is wasted power T or <u>F</u>
9. Reactance is measured in KVAR <u>T</u> or F
10. True Power is measured in KW <u>T</u> or F
11. Apparent Power is measured in KVA <u>T</u> or F

Power Factor Correction - Sizing Location and Benefits

How to Determine if Your Facility Would Benefit from PF Study?

❑ What is Existing PF?

❑ Will adding Capacitors Improve PF?

❑ Will Correcting PF save $ and not inhibit Production?

Helpful Tip

To help you improve the power factor of motor driven systems consider the following:

- ✓ How to tell if your plant could benefit from capacitors.
- ✓ How to select capacitor schemes to eliminate power factor penalties and minimize losses.
- ✓ How to perform detailed plant surveys to collect sufficient data to determine where to put capacitors.
- ✓ Why the power system must be built with extra capacity to supply power.
- ✓ How reactive power contributes to additional losses.
- ✓ How capacitors, synchronous machines, and static (adaptive) power compensators correct for power factor.
- ✓ When to use switched versus fixed capacitors.
- ✓ When to use filtered capacitors and when not.
- ✓ How and when capacitors contribute to harmonic distortion problems.
- ✓ How to predict capacitors contribute to harmonic distortion problems.
- ✓ How capacitors can fall prey to harmonics and switching transients.

Sizing and Locating Power Factor Correction Capacitors

Once you decide that your facility can benefit from power factor correction, you will need to choose the optimum type, size, number, and strategic locations for capacitors in your plant. The unit for rating power factor capacitors is the kVAR, which is equal to 1,000 volt-amperes of reactive power.

The kVAR rating signifies how much reactive power a capacitor will provide.

Industries That Typically Need PF Correction

Industry	Uncorrected Power Factor
Saw Mills	45% – 60%
Plastics (extruders)	55% – 70%
Machine Shops	40% – 65%
Plating, textiles, chemicals, breweries	65% – 75%
Foundries	50% – 80%
Chemicals	65% – 75%
Textiles	65% – 75%
Arc Welding	35% – 60%
Cement Works	78% – 80%
Printing	55% – 70%

The value of individual motor reactive power is cumulative toward the overall plant reactive power. Therefore, when you improve the power factor of a single motor, you are reducing the plant's reactive power requirement.

Although power factor correction capacitors reduce current in the lines supplying the motor, they do not reduce motor current, input power requirements, or performance. The greatest power factor correction benefits are obtained by placing capacitors at the source of reactive currents.

It is common to distribute capacitors on motors throughout an industrial plant.

This is a good strategy when capacitors must be switched to follow a changing load.

The advantages of bank installations downstream of the utility meter at the plant substation or service entry include the following:

• The cost per kVAR is lower.
• Installation costs are also lower.
• The total plant power factor improves, which reduces or eliminates utility power factor penalty charges.
• Total kVAR may be reduced because all capacitors are on line even when some motors are switched off.
• Automatic switching ensures the exact amount of power factor correction and eliminates over capacitance and resulting overvoltages.

If your facility operates at a constant load around the clock, fixed capacitors are the best solution. If the load is variable, such as two 7-hour production shifts, followed by a nighttime cleanup shift 5 days per week, you will need switched units to decrease capacitance during times of reduced load.

If your feeders or transformers are overloaded, or if you wish to add additional load to already loaded lines, you should apply power factor correction at the load. If your facility has excess current-carrying capacity, you can install capacitor banks at main feeders. There are three location options (A, B, & C) for installing capacitors at the motor.

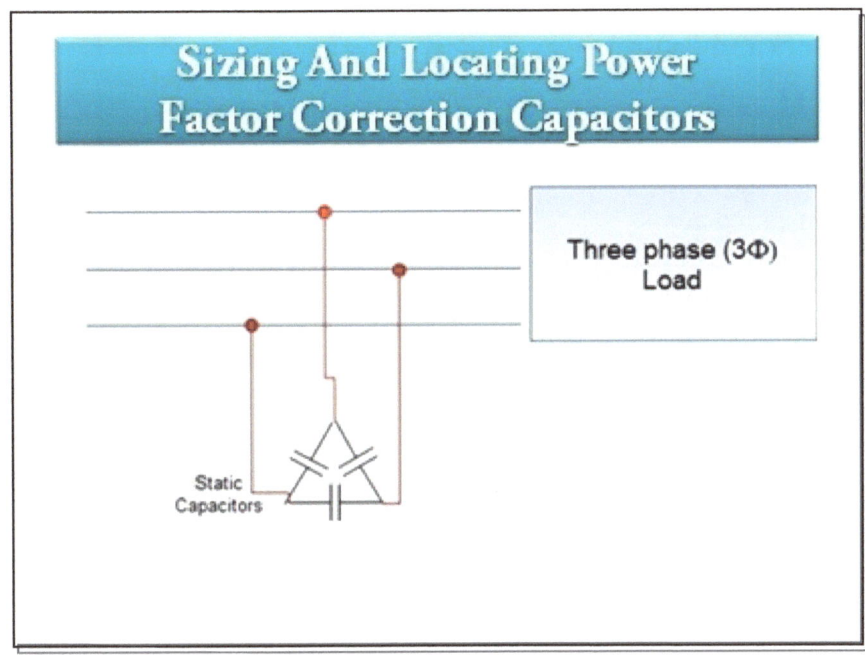

These options, along with the types of motors applicable to each, are as follows:

Location A — At the motor side of the overload relay
• New motor installations, in which overloads can be sized in accordance with a lower current draw
• Existing motors, when no overload change is required

Location B — Between the starter and overload relay
• Existing motors, when placement at location A would allow overload current to surpass code

Location C — At the line side of the starter
• Motors that are jogged or reversed (jogging refers to service conditions that include repeated starting and stopping of a motor such as moving a crane or a conveyor to a particular location).
• Multispeed motors
• Starters with open transition and those that disconnect/reconnect the capacitor during cycle
• Motors that start frequently
• Motor loads with high inertia.

Sizing Capacitors for Individual Motors and entire plant Loads.
Capacitors which are installed across the motor terminals and switched with the motor should not be sized larger than the amount of kVAR necessary to raise the motor no-load power factor to 100%.

 Capacitor manufactures and apps available to size capacitors.

Sizing Capacitors for Individual Motors and Entire Plant Loads

Capacitors which are installed across the motor terminals and switched with the motor should not be sized larger than the amount of kVAR necessary to raise the motor no-load power factor to 100%.

Always size capacitors for individual motor loads based on motor frame, synchronous speed (RPM), and horsepower.

Capacitors size are measured based the kVAR necessary to correct the power factor to 95%.

If you know the total plant load (kW), your present power factor, and the power factor you intend to achieve, size capacitors based on the KVAR.

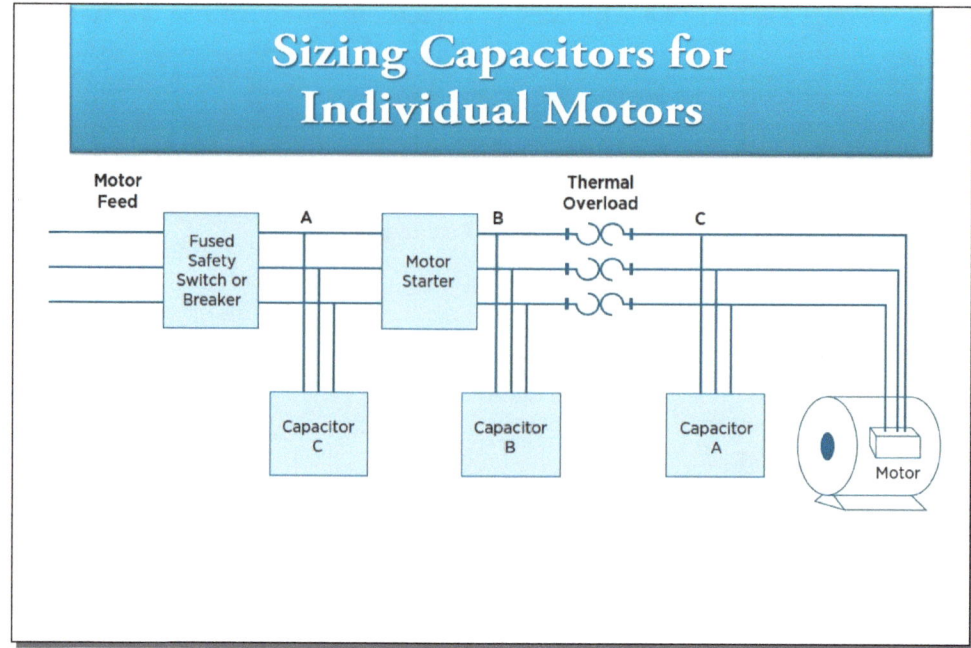

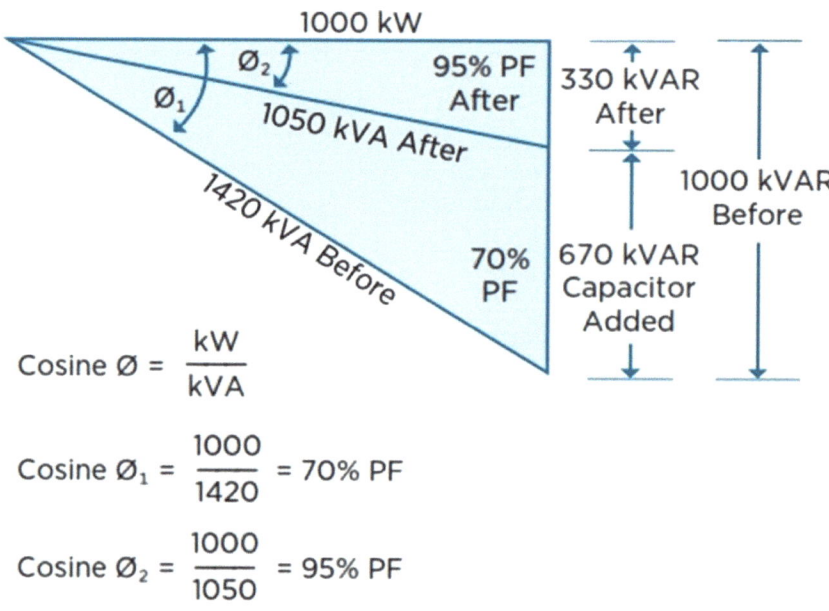

$$\text{Cosine } \emptyset = \frac{kW}{kVA}$$

$$\text{Cosine } \emptyset_1 = \frac{1000}{1420} = 70\% \text{ PF}$$

$$\text{Cosine } \emptyset_2 = \frac{1000}{1050} = 95\% \text{ PF}$$

The table (next page) is useful for sizing banks of capacitors that can be located at motor control centers, feeders, branch circuits, or the plant service entrance.

The power triangle indicates the demands on a plant distribution system before and after adding capacitors to improve power factor.

Increasing the power factor from 70% to 95% reduces the apparent power from 1,420 kVA to 1,050 kVA, *a reduction of 26%.*

Motor Capacitor Sizing Guide

NEMA Code	B																		C	D	Wound Rotor
	Before 1955						U-Frame						T-Frame								
Poles	2	4	6	8	10	12	2	4	6	8	10	12	2	4	6	8	10	12	4-6	8	6	
RPM	3600	1800	1200	900	720	600	3600	1800	1200	900	720	600	3600	1800	1200	900	720	600	1800 1200	900	1200	
HP=3	1.5	1.5	1.5	2	2.5	3.5	1	1	1	2			1.5	1.5	2.5	3	3	4				
5	2	2	2	3	4	4.5	1	2	2	2			2	2.5	3	4	4	5				
7.5	2.5	2.5	3	4	5.5	6	1	2	4	4			2.5	3	4	5	5	6				
10	3	3	3.5	5	6.5	7.5	2	2	4	5	5	5	4	4	5	6	7.5	8				
15	4	4	5	6.5	8	9.5	4	4	4	5	5	5	5	5	6	7.5	8	19	5	5	5	5.5
20	5	5	6.5	7.5	9	12	4	5	5	5	10	10	6	6	7.5	9	10	12	5	6	6	7
25	6	6	7.5	9	11	14	5	5	5	5	10	10	7.5	7.5	8	10	12	18	6	6	6	7
30	7	7	9	10	12	16	5	5	5	10	10	10	8	8	10	14	15	23	7.5	9	10	11
40	9	9	11	12	15	20	5	10	10	10	10	15	12	13	16	18	23	25	10	12	12	13
50	12	11	13	15	19	24	5	10	10	15	15	20	15	18	20	23	24	30	12	15	15	18
60	14	14	15	18	22	27	10	10	10	15	20	25	18	21	23	26	30	35	18	18	18	20
75	17	16	18	21	26	33	15	15	15	20	25	30	20	23	25	28	33	40	19	23	23	25
100	22	21	25	27	33	40	15	20	25	25	40	45	23	30	30	35	40	45	27	27	30	33
125	27	26	30	33	40	48	20	25	30	30	45	45	25	36	35	42	45	50	35	38	38	40
150	33	30	35	38	48	53	25	30	30	40	45	50	30	42	40	53	53	60	38	45	45	50
200	40	38	43	48	60	65	35	40	60	55	55	60	35	50	50	65	68	90	45	60	60	65

These tables are available from most capacitor manufactures. Several calculation apps are also available to size capacitors.

Multipliers to Determine Capacitor kVAR Required for Power Factor Correction

Original Power Factor	Corrected Power Factor																		
	0.80	0.81	0.82	0.83	0.84	0.85	0.86	0.87	0.88	0.89	0.90	0.91	0.92	0.93	0.94	0.95	0.96	0.99	1.0
0.50	0.982	1.008	1.034	1.060	1.086	1.112	1.139	1.165	1.192	1.220	1.248	1.276	1.306	1.337	1.369	1.403	1.440	1.589	1.732
		0.962		1.015		1.067		1.120			1.203		1.261		1.324		1.395		
		0.266	.294		0.344	.372		0.4..	.452		0.5..	.536		0...	.629		0...	0.845	.2
0.72	0.214	0.240	0.266	0.292	0.318	0.344	0.371	0.397	0.424	0.452	0.480	0.508	0.538	0.569	0.601	0.635	0.672	0.821	0.964
0.73	0.186	0.212	0.238	0.264	0.290	0.316	0.343	0.369	0.396	0.424	0.452	0.480	0.510	0.541	0.573	0.607	0.644	0.793	0.936
0.74	0.159	0.185	0.211	0.237	0.263	0.289	0.316	0.342	0.369	0.397	0.425	0.453	0.483	0.514	0.546	0.580	0.617	0.766	0.909
0.75	0.132	0.158	0.184	0.210	0.236	0.262	0.289	0.315	0.342	0.370	0.398	0.426	0.456	0.487	0.519	0.553	0.590	0.739	0.882

Instructions:
1. Find the present power factor in column.
2. Read across to optimum power factor column.
3. Multiply that number by kW demand.

Example: If your plant consumed 410 kW, was currently operating at 73% power factor and you wanted to correct power factor to be 95% you would.
1. Find 0.73 in column.
2. Read across to 0.95 column.
3. Multiply 0.607 by 410 = 249 (round to 250.) 4. You need 250 kVAR to bring your plant to 95% power factor.

If you don't know the existing power factor level of your plant, you will need to calculate it before using this table.

To calculate existing power factor:
kW divided by kVA = Power Factor

Problem and Solution

Problem –
- Plant consumption 410 kW @ 73% power factor
- Desired Correction of Power Factor is 95%

Solution - using table in previous slide
- Find 0.73 in vertical column Original Power Factor
- Go to 0.95 horizontal column
- Multiply 0.607 by 410 = 249 (round to 250.)
- Answer 250KVAR to bring plant to 95%

To Solve Use Previous Table:
1. Find the vertical column <u>original power factor</u>.
2. Read across horizontal column.
3. Multiply 0.607 by 410 = 249 (round to 250.)
4. Solution - You need 250 kVAR to bring your plant to 95% power factor If you don't know the existing power factor level of your plant, you will have to calculate it before using this table.

To Calculate Existing Power Factor:

kW divided by kVA = Power Factor

Utility Rate Calculations

In this scenario, the utility charges according to kW demand ($4.50/kW) and include a surcharge or adjustment for low power factor. The following formula shows a billing adjustment based upon a desired 95% power factor.

Plant Conditions:

For our sample facility, the original demand is 4,600 kVA × 0.80, or 3,680 kW. The multiplier applies to power factors up to 0.95. Billing Before power Factor Correction 3,680 kW × 0.95 ÷0.80
= 4,370 × $4.50
= $19,665/month or $235,980/year
Where: kW Billing after power Factor Corrected to 95% Savings are $37,260/year.
kW BILLED DEMAND BILLED
= kW DEMAND 3,680 kW × 0.95 0.95
= 3,680 × $4.50
= $16,560/month or $198,720/year × 0.95
= Adjusted or billable demand
= Measured electric demand in kW
PF = Power factor as a decimal

Cost Without PF Correction

- Facility Demand = 4,600 kVA × 0.80, or 3,680 kW
- Billing Before Power Factor Correction
  - 3,680 kW × 0.95 ÷ 0.80 =
  - 4,370 × $4.50 =
  - $19,665/month or $235,980/year

Note: billing adjustment based upon a desired 95% power factor

---

Cost With PF Correction

- Facility Demand = 4,600 kVA × 0.80, or 3,680 kW
- Billing After Power Factor Correction to 95%
  - 3,680 kW × 0.95 ÷ .95 = 3680 kW
  - 3,680 kW × $4.50 = $16,560
  - $16560/month or $198720 /year
  - Savings $37,260.00

Note: billing adjustment based upon a desired 95% power factor

Return on Investment – two years

Cost to install and PF Correction Less than $60,000

Power Factor Correction Costs

Average Cost for Capacitors on a 480-volt system is:

✓ **Capacitor Cost Average**

    ✓ **$30 without harmonic filtering per kVAR**

    ✓ **$55 with harmonic filtering per kVAR**

✓ **Installation Cost: $30 - $40 per KVAR**

✓ **NOTE 1: Costs based on Capacitors exceeding 100 KVAR**

✓ **NOTE 2: Capacitor Bank is lower installation costs than individual capacitors at motors**

Contact Info:

If you have comments or questions feel free to contact Dr. Carpenter:

David@integrityco.com

Look at website for many other books, online classes and videos at www.integrityco.com

www.ingramcontent.com/pod-product-compliance
Lightning Source LLC
Chambersburg PA
CBHW040746200526
45159CB00023B/1744